ELEMENTS OF THE PERIODIC TABLE

WHAT IS NITROGEN?

007

N

NITROGEN

14.01

I0807251

KATHLEEN A. KLATTE

ROSEN PUBLISHING

Published in 2026 by The Rosen Publishing Group, Inc.
2544 Clinton Street, Buffalo, NY 14224

Copyright © 2026 by The Rosen Publishing Group, Inc.

Portions of this work were originally authored by Heather Hasan and published as *Nitrogen*. All new material this edition authored by Kathleen A. Klatte.

Designer: Rachel Rising
Editor: Kathleen A. Klatte

All rights reserved. No part of this book may be reproduced in any form without permission in writing from the publisher, except by a reviewer.

Cataloging-in-Publication Data
Names: Klatte, Kathleen A.
Title: What is nitrogen? / Kathleen A. Klatte.
Description: Buffalo, NY : Rosen Publishing, 2026. | Series: Elements of the Periodic Table | Includes glossary and index.
Identifiers: ISBN 9781499478457 (pbk.) | ISBN 9781499478464 (library bound) | ISBN 9781499478471 (ebook)
Subjects: LCSH: Nitrogen--Juvenile literature. | Group 15 elements--Juvenile literature. | Chemical elements--Juvenile literature.
Classification: LCC QD181.N1 K538 2026 | DDC 546'.711--dc23

Some of the images in this book illustrate individuals who are models. The depictions do not imply actual situations or events.

Manufactured in the United States of America

CPSIA Compliance Information: Batch #CSRYA26. For further information, contact Rosen Publishing at 1-800-237-9932.

Find us on

CONTENTS

INTRODUCTION

Most people today associate Alfred Nobel (1833–1896) with the prizes awarded each year in his honor—most notably the Nobel Peace Prize. However, during his lifetime, he was known as a chemist, engineer, and inventor. He held patents for 355 inventions and was able to amass a large personal fortune.

As a young man, Nobel studied chemistry in Paris, France. After that, he studied in the United States with the engineer who built the ironclad ship *Monitor*. He then returned to work in his father's factory in Russia. The factory produced military equipment for the Crimean War.

Nobel was very interested in explosives and making them safer to use in mining. In the 1860s, he began experimenting with nitroglycerin. Nitroglycerin was first made by an Italian chemist, Ascanio Sobrero (1812–1888), in the 1840s. It's a liquid made by mixing glycerol (an oily, syrupy liquid) with nitric acid (containing nitrogen) and sulfuric acid. Nitroglycerin is a very powerful explosive—a small amount can cause enormous damage. In liquid form, it's also highly unstable. Nitroglycerin needs to be handled with the utmost care to avoid accidental explosions, such as the one that killed Nobel's brother in 1864.

That same year, Nobel patented a detonator, or blasting cap, so nitroglycerin could be used in a more controlled manner. In 1866, following a near disaster in his lab, Nobel discovered that nitroglycerin is much more stable when mixed with less sensitive things. He mixed nitroglycerin with a form of rock known as diatomaceous earth. The resulting solid form was much safer to transport. He named it dynamite, from the Greek word *dynamis*, which means power.

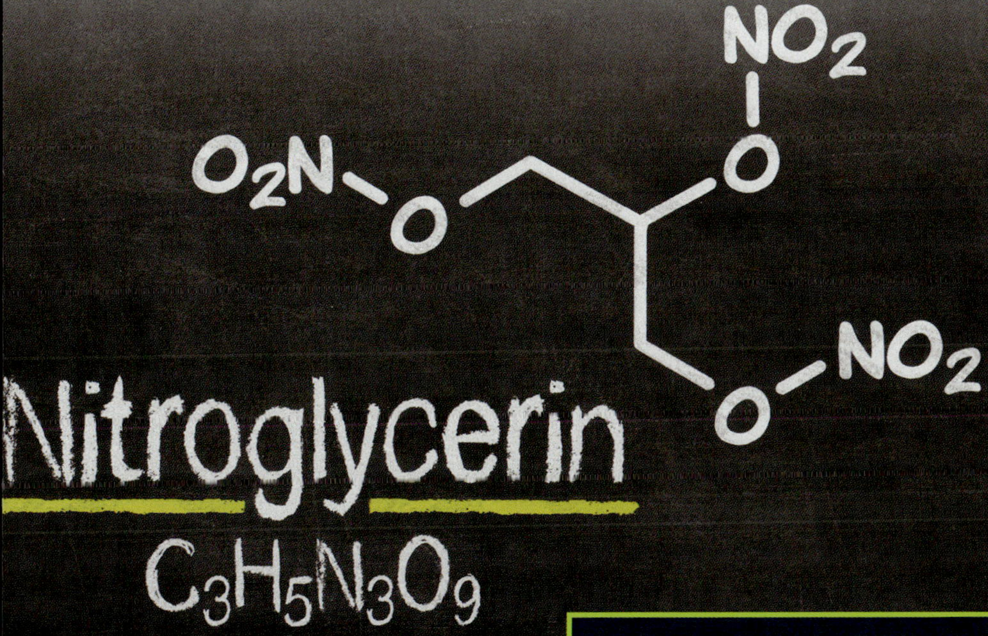

Nitroglycerin is a chemical compound made up of carbon, hydrogen, nitrogen, and oxygen. This substance can end lives explosively—or save them.

Nitrogen's spot on the periodic table has lots of information--its name, chemical symbol, atomic number, and atomic weight.

Although Alfred Nobel intended for dynamite to be used to make mining safer, it, along with blasting caps and other inventions, are often used for war and violence. He regretted this and decided to leave much of his fortune for a set of prizes awarded each year for extraordinary accomplishments in physics, chemistry, medicine, literature, and efforts toward world peace.

Ascanio Sobrero also regretted his part in discovering nitroglycerin. He would probably be happy to know that today, it also has another use—as a powerful medicine to help heart patients. Today, doctors may prescribe nitroglycerin as a tablet that a patient places under their tongue to help with chest pain or at the first sign of a heart attack. Inside the human body, nitroglycerin is converted to nitric oxide. This relaxes the muscles that form blood vessels, allowing them to dilate. This lowers blood pressure. It's also prescribed in other forms.

Nitroglycerin isn't the only form of nitrogen. Nitrogen is one of the most abundant elements on Earth. We interact with it in the food we eat and the air we breathe. Read on to find out more about this exciting element!

Today, Alfred Nobel's legacy lives on in awards granted for achievements in science, literature, and world peace

CHAPTER 1

NECESSARY NITROGEN

Nitrogen (N) has been known by different names over the centuries—"noxious air," "foul air," and azote (derived from a Greek word meaning "without life")—all because pure nitrogen gas can't support life. It's odorless, tasteless, and colorless. However, nitrogen is all around us, binding with other elements into forms necessary for life.

Nitrogen makes up most of our atmosphere—about four-fifths. It's found in the food we eat, the fertilizer we use to grow crops, and even in our bodies. Life on Earth needs nitrogen to survive.

Like many other things, nitrogen isn't good or bad on its own. It can be combined with other elements to form life-saving medicines. In different proportions, the same combination of elements can also form deadly explosives. Liquid nitrogen is used in refrigeration systems and in medicine to preserve blood. Nitrogen gas is used to preserve food and as a propellant for aerosols. However, this useful gas is also a component of many forms of pollution.

BY ANY OTHER NAME

Since pure nitrogen gas does not do very much, it took a long time for people to notice it. Scientists didn't discover nitrogen until the late 18th century. At that time, they knew that air was made up of at least two gases, and many scientists were working hard to determine what these gases were. Though several scientists were on the road to discovering nitrogen around the same time, credit for its discovery was given to a Scottish scientist named Daniel Rutherford (1749–1819). In 1772, Rutherford announced his discovery of what he called "noxious air." He named it this because animals could not breathe it nor could candles burn in a container filled with it.

Nitrogen is one of the most important elements for healthy soil and plants. It needs to be replaced in the soil to maintain healthy levels for growing plants. One of the oldest methods of enriching soil is with composted manure, or animal waste.

However, it was not until 1775 that a French scientist named Antoine Lavoisier (1743–1794) suggested that this "noxious air" was actually an element. It was also Lavoisier who later suggested the name "azote," which is what the French still call nitrogen today. The term "nitrogen" was first used by Jean-Antoine-Claude Chaptal (1756–1832) in 1790. This name was created from the French words *nitre* and *gène*, which together mean "niter forming." Niter is the common name for potassium nitrate, a naturally occurring salt that was a known source of nitrogen at the time.

AN INVITATION TO THE TABLE

All the materials and objects around you are made up of elements or combinations of elements. There are more than 100 known elements today. To organize these elements, scientists arranged them on a big chart called the periodic table. The periodic table that we use today is based on the work of a Russian chemist named Dmitri Mendeleev (1834–1907). Mendeleev published the first version of the table in 1869. He came up with the idea while teaching chemistry at the University of St. Petersburg in Russia. Mendeleev sought to organize the elements in a way that would make it easier for his students to study and understand them. He arranged the elements in horizontal rows according to weight, with the lightest element of each row on the left end and the heaviest one on the right. Though Mendeleev's periodic table did not list all the elements that we know of today, nitrogen was among those that he included on his first chart.

WHAT ARE WE BREATHING?

In 1772, Swedish chemist Carl Wilhelm Scheele (1742–1786) proved that air was composed of at least two different gases. He called these "fire air" and "foul air." Fire air was air that burned—oxygen, which Scheele would share the credit for discovering. Foul air was what remained after combustion—nitrogen. However, Daniel Rutherford published his findings first and is credited with the discovery of nitrogen.

Today, with modern measuring equipment, scientists know that while air is mainly composed of oxygen and nitrogen, it also contains traces of a few other elements:

NITROGEN (N_2): 78 percent

OXYGEN (O_2): 21 percent

ARGON (AR): 0.9 percent

CARBON DIOXIDE (CO_2): 0.04 percent

HYDROGEN (H_2), HELIUM (HE), KRYPTON (KR), NEON (NE), XENON (XE), AND OTHER ELEMENTS AND PARTICLES: 0.06 percent collectively

Trees replenish Earth's oxygen supplies via the process of photosynthesis.

THE LATEST ADDITIONS

Mendeleev realized that there were probably more elements in the world than had been discovered. He left spaces in the periodic table for new elements to be added. His theory was proven correct, as more elements were discovered and added to the table during his lifetime. Improvements in technology have led to the discovery and creation of additional elements. New discoveries are still being made today. The latest additions to the periodic table are four superheavy elements created in laboratories. In 2016, elements 113, 115, 117, and 118 were added to the table.

ATOMIC NUMBER	INITIAL DESIGNATION	OFFICIAL NAME
113	UNUNTRIUM (UUT)	NIHONIUM (NH)
115	UNUNPENTIUM (UUP)	MOSCOVIUM (MC)
117	UNUNSEPTIUM (UUS)	TENNESSINE (TS)
118	UNUNOCTIUM (UUO)	OGANESSON (OG)

Oganesson, the latest addition, is identified by the letters Og, the atomic number 118, and the atomic weight 294. The atomic number is the number of protons contained in the atom's nucleus. The atomic weight is the sum of protons and neutrons (plus a little from electrons) in the atom's nucleus.

118

Og

Oganesson

(294)

CHAPTER 2

IT'S ELEMENTAL

An element is composed of only one type of atom. It can't be broken down any further. Every atom of an element (such as nitrogen) is the same. For a long time, scientists thought that atoms were the smallest particles. Then in 1897, the first subatomic, or smaller than an atom, particles—electrons—were discovered. Protons and neutrons were discovered in the 1930s and quarks in the 1970s. Now, scientists define an atom as a unique arrangement of subatomic particles.

Electrons orbit the nucleus of an atom in shells. This arrangement is what makes it possible for elements to share electrons and form compounds.

NITROGEN SNAPSHOT

CHEMICAL SYMBOL: N

DISCOVERED BY: Daniel Rutherford in 1772

CLASSIFICATION: Nonmetal

PROPERTIES: Nonreactive, nonflammable, colorless, odorless

ATOMIC NUMBER: 7

ATOMIC WEIGHT: 14.00674

PROTONS: 7 **ELECTRONS**: 7 **NEUTRONS:** 7

DENSITY AT 293 K: 0.0012506 g/cm³

FREEZING/MELTING POINT: -345.82°F; -209.9°C; 63.25 K

BOILING POINT: -320.44°F; -195.8°C; 77.35 K

COMMONLY FOUND: Air, soil, stars

Today's version of Mendeleev's table is found in chemistry textbooks and classrooms around the world.

WHAT'S SMALLER THAN AN ATOM?

The three main subatomic particles that make up an atom are neutrons, protons, and electrons. Neutrons and protons are clustered together at the center of the atom to form a dense core called the nucleus. Neutrons carry no electrical charge, while protons have a positive electrical charge. This gives the nucleus an overall positive electrical charge. Nitrogen has seven protons in its nucleus, so its nucleus has a charge of +7.

Around the nucleus of an atom are negatively charged electrons, arranged in layers, or shells. The electrons are not fixed in a single position but orbit, or circle, the nucleus. The negative electrons are attracted to the positive nucleus, and it is this attraction that holds the electrons around the nucleus. The electrons are not pulled all the way into the nucleus because the negatively charged electrons repel each other. The number of protons and electrons is equal, so the positive and negative charges of the atom balance. Therefore, since nitrogen has seven protons, it also has seven electrons.

BY THE NUMBERS

What makes nitrogen different from other elements such as oxygen or silver? The difference lies in the number of protons that are found in the nucleus of each atom. Since it is the number of protons that makes each element unique, it makes sense that the elements on the periodic table are organized by these numbers.

The number of protons in one atom of an element is called the atomic number. On the periodic table, this number is found above and to the left of an element's symbol. Since an atom of nitrogen has seven protons, its atomic number is 7. The fact that nitrogen has seven protons in its nucleus is what makes it nitrogen. Let's imagine that we could add a proton to nitrogen's nucleus (which is physically impossible). Now its nucleus would no longer have seven protons; it would have eight. We would now have an atom of oxygen. We cannot breathe without oxygen. Just one proton makes the difference between an element that allows us to breathe and one that doesn't. If we could take away one of nitrogen's protons, we would have carbon (C), which has six protons in its nucleus. Carbon is very different from nitrogen. While nitrogen is a colorless gas, we see carbon in the form of diamonds and coal or the graphite in your pencil. As you can see, changing one little proton can make a world of difference!

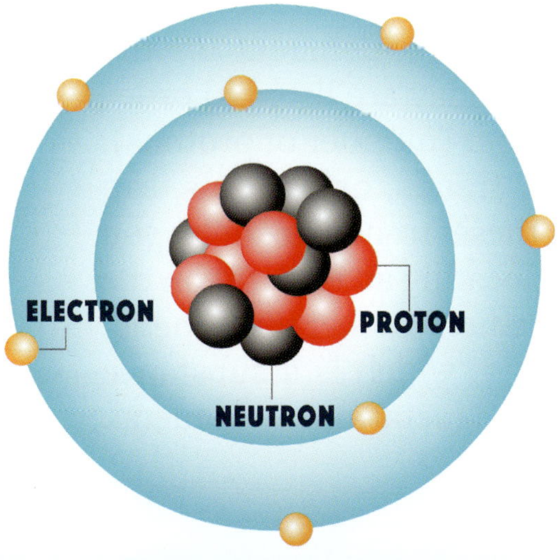

In this model you can see seven protons and seven neutrons in the nucleus of a nitrogen atom. It also has two shells with a total of seven electrons orbiting the nucleus. The outer shell is called the valence. Valence electrons are the ones that have contact with atoms of other elements, forming compounds and causing chemical reactions.

WHAT DOES AN ATOM WEIGH?

Another number found on an element's square on the periodic table is called the atomic weight (also called atomic mass). The approximate atomic weight is the sum of the number of protons and neutrons in the atom. Nitrogen has an atomic weight of 14. When we know the atomic weight of an element, it helps us to figure out how many neutrons there are in an atom of that element. Knowing that the atomic weight of nitrogen is 14 and the atomic number (number of protons) is 7, we can figure out how many neutrons there are in an atom of nitrogen by subtracting the two numbers:

ATOMIC WEIGHT - ATOMIC NUMBER = NUMBER OF NEUTRONS

14 - 7 = 7

As you can see, nitrogen has seven neutrons in its nucleus. We can find out a lot of information about nitrogen just by looking at the periodic table!

A TRENDY TABLE

Unlike Mendeleev's chart, the elements on the periodic table that we use today are listed in order of increasing atomic number (number of protons). Arranged like this, many trends, or patterns, can be seen. You can use these trends to help you classify the elements. By locating an element on the periodic table, you can predict whether it is a metal, a nonmetal, or a metalloid, which is a substance with characteristics of both a metal and a nonmetal.

If you look at the periodic table, you will notice that the elements are divided by a "staircase" line. The metals are found to the left of this line, and the nonmetals are found to the right. Most of the elements bordering the line are metalloids, or semimetals. These act like metals sometimes and other times like nonmetals.

Nitrogen is part of a group of elements called pnictogens, a word derived from Greek words that translate roughly to "choking gas." The group also includes phosphorus (P), arsenic (As), antimony (Sb), bismuth (Bi), and moscovium (Mc).

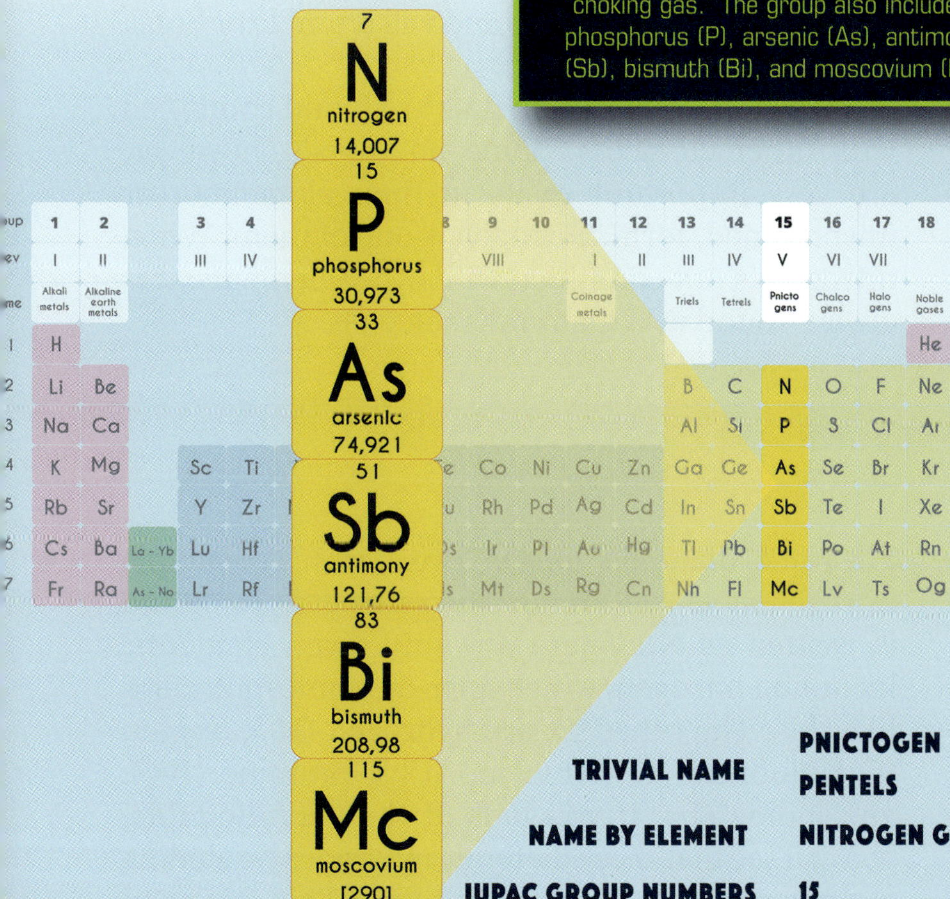

TRIVIAL NAME	PNICTOGEN PENTELS
NAME BY ELEMENT	NITROGEN GROUP
IUPAC GROUP NUMBERS	15

Nitrogen is found to the right of the staircase line. This is where you would expect to find it, since nitrogen is a gas. A nonmetal is an element that does not have the characteristics of a metal. Metals are shiny, flexible, malleable (they can be pounded into sheets), and ductile (they can be pulled into wires). They are also good conductors of electricity and heat. By contrast, nonmetals are not shiny. They also do not conduct heat or electricity well. Nearly half of the nonmetals are colorless gases, but they can also be liquids and solids. However, unlike solid metals, solid nonmetals are brittle and will crumble or break if pulled or hammered.

Many times, nonmetals are found combined with other elements. Some nonmetals, however, are very important to us in their elemental, or uncombined, form. Nitrogen is one of them. When nitrogen is not combined with other elements, it is always found as a diatomic molecule.

TWO ATOMS ARE BETTER THAN ONE

The word "diatomic" means "two atoms." A diatomic nitrogen molecule contains two nitrogen atoms and is written as N_2. There are only seven elements, including nitrogen, which form diatomic molecules like this. The other six are hydrogen (H_2), oxygen (O_2), fluorine (F_2), chlorine (Cl_2), bromine (Br_2), and iodine (I_2). If you look at the periodic table, you can see that these diatomic molecules (excluding hydrogen) form the number seven. This is a good way for you to remember these special elements.

FINDING YOUR PLACE

The periodic table is very useful for many reasons. An element's location on the table will tell you a lot about the element. As you look across the table from left to right, the horizontal row of elements is called a period. Elements are arranged in periods by the number of electron shells that surround the nucleus of their atoms. Nitrogen is in period 2. Therefore, the electrons in a nitrogen atom occupy two shells around its nucleus.

A diatomic nitrogen molecule is formed when two nitrogen atoms are connected by a triple bond.

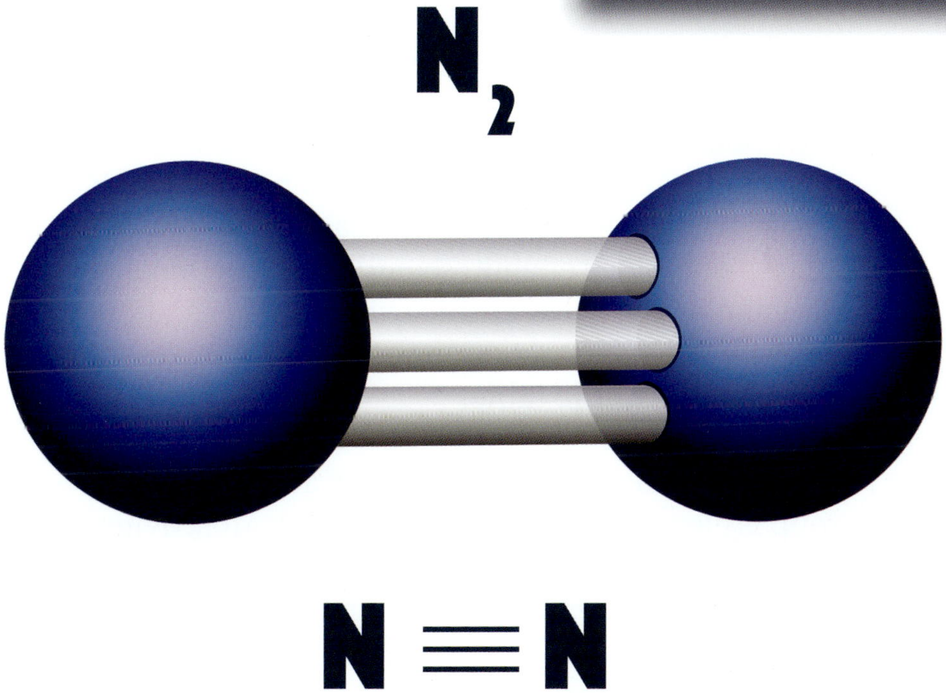

As you read down the chart from top to bottom, the column of elements you see is called a group or family. Just as you might have similar characteristics to the other members of your family, the elements in a group have similar properties. Nitrogen heads the family of elements that make up one of these groups. Depending on how old the version of the periodic table you're looking at is, the group might be called VA or 5A. However, the 2022 International Union of Practical and Applied Chemistry (IUPAC) update of the periodic table calls this family 15. Changes happen because science is constantly evolving as new technology enables scientists to learn more about the way the world works.

ALL IN THE FAMILY

The International Union of Practical and Applied Chemistry is the global chemistry organization that updates and publishes the periodic table of elements. It approves the addition of new elements as scientists discover them or create them in a lab. As new technology enables scientists to view smaller and smaller particles and learn new things about chemical elements, sometimes it reorganizes or relabels parts of the table.

There are currently six elements in the nitrogen family. The most recent addition to the group is element 115, moscovium (Mc). Moscovium is a superheavy element that does not occur in nature. It was created in labs and initially called ununpentium. It was placed in the nitrogen family because its structure is similar to other elements in the group. In 2016, after many years of research, the element was given its proper name and place on the table.

All six of the elements in the nitrogen group have five electrons in their outermost shell. These outermost electrons are called valence electrons, and they determine how an element acts. Because they all have the same number of electrons in their outermost electron shell, all the elements in the group have similar chemical properties, even though nitrogen is one of the lightest elements and muscovium is one of the heaviest.

Arsenic is part of the nitrogen group. In the 19th century, people used it to create brilliant green dyes used for clothing, book binding (shown here), paints, and wallpaper. But arsenic is highly toxic, and these dyes made people sick—especially people who worked with the dyes daily in factories.

CHAPTER 3

ALL ABOUT NITROGEN

Each element of the periodic table has properties that make it different from every other element. These properties can be physical or chemical. Physical properties are things that can be observed or recorded without changing the substance into a different substance.

Physical properties include things like:

- boiling point
- freezing point
- melting point
- color
- density

Frozen nitrogen is still nitrogen. Changing the temperature hasn't changed its chemical composition. A chemical change is one that alters the identity of an element through bonding or reacting with other elements. For example, nitrogen reacts with hydrogen to form ammonia, a different substance.

Liquid and frozen nitrogen are dangerously cold and require special handling.

IT'S A GAS!

At room temperature, an element is found in one of three physical states: solid, liquid, or gas. Knowing the physical state, or phase, of an element at room temperature helps scientists to identify it. Nitrogen differs from the rest of the elements in its family because, at room temperature, nitrogen is a gas.

Gases, such as nitrogen, do not have definite shapes or volumes. If put in a container, a gas will take the shape of the container. Gases differ from solids and liquids because they can be compressed and are able to expand. Gases will fit in a container of almost any shape and size. If put in a large container, a gas will expand to fill it. It can also be compressed to fit into a smaller container. If not confined to a container, gases will disperse into space.

Liquid nitrogen is very useful in medicine and science. It's used to preserve and transport tissues and medicine.

Liquid nitrogen is used in the food industry to create cool smoke effects. However, the "smoke" must be allowed to evaporate or diners risk frostbite.

HOW COOL IS THAT?

In nature, nitrogen is always a gas. However, if this gas is cooled to a low enough temperature, nitrogen can become a liquid or even a solid. For nitrogen to condense, or turn into a liquid, the temperature must be colder than -320.44°F (-195.8°C). That's cold! As a liquid, nitrogen is colorless and odorless, resembling ordinary water.

For nitrogen to reach its freezing point (the temperature at which it becomes a solid), the temperature must drop even further to -345.82°F (-209.9°C). Not even in the coldest places on Earth do the temperatures drop that low! In fact, the coldest temperature ever recorded on Earth is -128.2°F (-89°C), nowhere near either of these points.

SOLID, LIQUID, OR GAS?

Density is how compact an object is, or how much mass it contains per unit of volume. Solids are denser than liquids, which are, in turn, denser than gases. Solid nitrogen has a density of 1.03 grams per cubic centimeter. This is just a little higher than the density of liquid nitrogen, which is 0.81 g/cm^3. In comparison, the density of water is 1 g/cm^3. Gaseous nitrogen has a much lower density than both solid and liquid nitrogen. Its density is only 0.00125 g/cm^3. These densities can give you an idea about how far apart the nitrogen atoms are in each phase. The atoms in solid nitrogen are packed very closely together. The atoms in liquid nitrogen are not as closely packed, but as with solid nitrogen, they still touch one another. However, the nitrogen atoms in gaseous nitrogen are quite far from one another. In fact, these atoms are so spread out that the volume of the gas is on average about 750 times greater than that of a sample of solid or liquid nitrogen of the same mass.

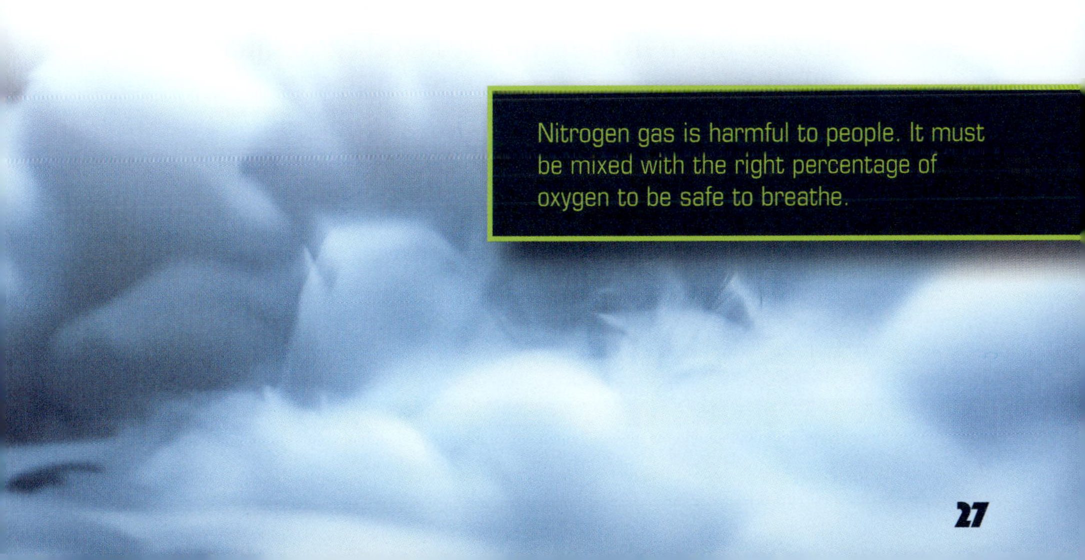

Nitrogen gas is harmful to people. It must be mixed with the right percentage of oxygen to be safe to breathe.

BETTER WITH TWO!

A nitrogen atom consists of a nucleus containing seven neutrons and seven protons. Seven electrons orbit the nucleus in two shells. The inner shell has two electrons and the outer, or valence, shell contains five. A nitrogen molecule consists of two nitrogen atoms bonded together. This forms a diatomic molecule. Because both atoms are the same element, it's called a homonuclear diatomic molecule.

When atoms share electrons, they form a covalent bond. Atoms in a covalent bond can share one, two, or three pairs of electrons. Because a nitrogen molecule shares three pairs of electrons (six shared electrons total), it forms a triple covalent bond. A triple covalent bond is strong and tight. This makes nitrogen molecules very stable. N_2 molecules don't react or mix with other elements. This means they are inert.

It takes a great deal of energy to break a nitrogen molecule apart so the two individual atoms can react with other elements. The energy bond of an N_2 molecule is 226 kilocalories per mole. Only carbon monoxide has a higher energy bond (256 kilocalories per mole).

What has enough energy to break apart a diatomic nitrogen molecule? A lightning strike!

This diagram clearly shows the arrangement of electrons in two shells orbiting each nitrogen atom. Three pairs of electrons in the valence shells form the strong triple covalent bond.

TRIPLE BONDS

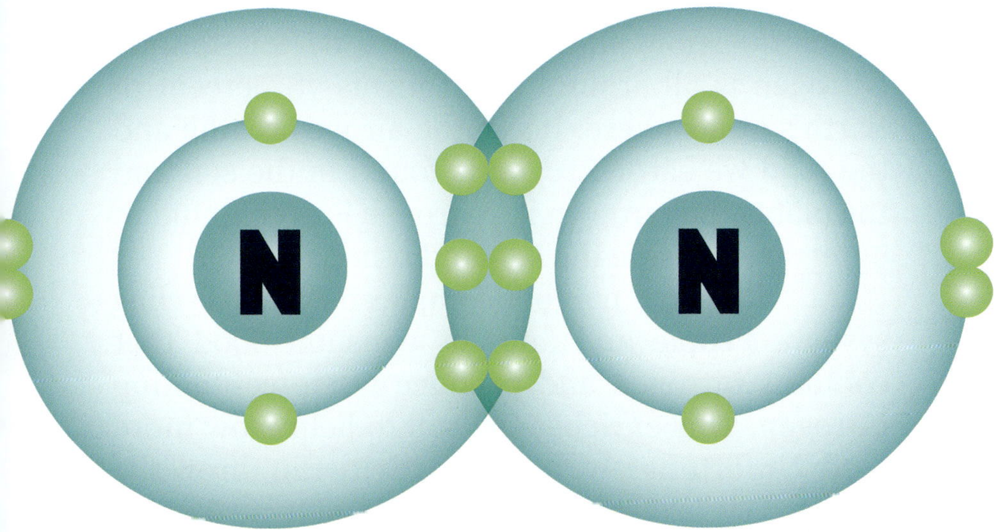

NITROGEN

$N_2 : N \equiv N$

SHARE 6 ELECTRONS

CHAPTER 4

HERE, THERE, AND EVERYWHERE

Where is nitrogen found in the natural world? A better question might be, where isn't it found? Nitrogen is all around us. It forms 78 percent of Earth's atmosphere. It's in the air we breathe and the food we eat. Nitrogen is found in the soil and the plants that grow in it. Nitrogen is even found in outer space inside the sun and other stars.

Nitrogen is part of a complex natural recycling system. People and animals consume food that contains nitrogen and produce waste containing nitrogen. This enriches the soil to help grow more food. Unfortunately, human activity can throw the nitrogen cycle out of balance. Too much nitrogen in the environment can form pollutants like ammonia and ozone.

When too much nitrogen accumulates in water, it can cause algae to grow. This harms the plants and animals that live in the water. It can also make the water unsafe for drinking.

IN THE AIR

There is a lot of nitrogen in the air, but before it can be used, it must first be separated from the other gases found in air. Most commonly, nitrogen is obtained from air in a process called cryogenic distillation. In this process, air is cooled until it becomes a liquid. Air liquefies at about -274°F (-170°C) and between 8 and 10 units of atmospheric pressure, and then it becomes a boiling liquid. Each of the elements in the air turns back into a gas as that element's boiling point is reached. Scientists know that nitrogen has a boiling point of -320.4°F (-195.8°C). At that temperature, nitrogen leaves the liquid air as a gas. Since none of the other elements in the liquid share nitrogen's boiling point, nitrogen can be separated from the other elements. Scientists wait until the temperature of the liquid air reaches this point, and they collect the nitrogen gas as it comes out. Once the pure nitrogen has been isolated from the mixture, it is ready to be used commercially in a gas or liquid form.

Most of the nitrogen in our atmosphere is found in the troposphere. That's the lowest layer and includes the air that we breathe.

Chefs and bartenders may use liquid nitrogen to produce spectacular displays with food and drinks.

IN THE FREEZER AND PANTRY

Liquid nitrogen is used often in the food industry. Liquid nitrogen is so cold that it allows foods to instantly reach temperatures far below freezing. Freezing things at such low temperatures is called cryogenic freezing. Foods like fruit and cheesecake are packaged and then sprayed with liquid nitrogen. Upon contact, the liquid absorbs the heat from the food and evaporates, leaving the food frozen. The food is then ready to be shipped or stored. Foods treated with liquid nitrogen freeze so quickly that the food does not have time to lose moisture or grow bacteria, as do foods that are frozen by a normal process. Because foods that are frozen by liquid nitrogen are not significantly damaged as they freeze, they turn out fresher and better tasting when thawed.

Gaseous nitrogen is also used in food packaging. That "air" inside your potato chip bag is not air at all. It is nitrogen. Here, nitrogen does more than just keep your chips from getting crushed. It also keeps them from going stale. If the bags were filled with regular air, the oxygen in the air would react with the fat in the chips and make them stale. Nitrogen does not react with the chips, so they stay fresh. They stay fresh, however, only until you open the bag. Once you open your bag of chips, the nitrogen gas escapes and is replaced by air. Then it is only a matter of time until your chips lose their crunch.

Cooking with liquid nitrogen requires special training to keep both the chef and diners safe.

IN THE HOSPITAL

The properties that make nitrogen useful for food storage are also valuable in medicine and science. Liquid nitrogen is used to freeze specimens quickly for storage or transportation. This is how blood donations are stored safely until they're needed. Tissue samples are preserved this way for future research, and viruses can be stored for use in manufacturing vaccines. Fertility clinics rely on this technology to store eggs and other biological materials.

Liquid nitrogen is also used for certain medical procedures. Extreme cold causes frostbite, which can kill human tissues. Doctors can use very small amounts of liquid nitrogen under controlled conditions to kill small amounts of diseased or unwanted tissue, such as warts or certain kinds of skin cancer.

Nitrogen gas is used as packaging for certain medications. It protects them from the moisture and oxygen present in the air. Nitrous oxide (N_2O) is a gaseous compound of nitrogen and oxygen. It's used as a mild anesthetic. It's more commonly known as laughing gas because it often makes people feel funny and can cause them to laugh.

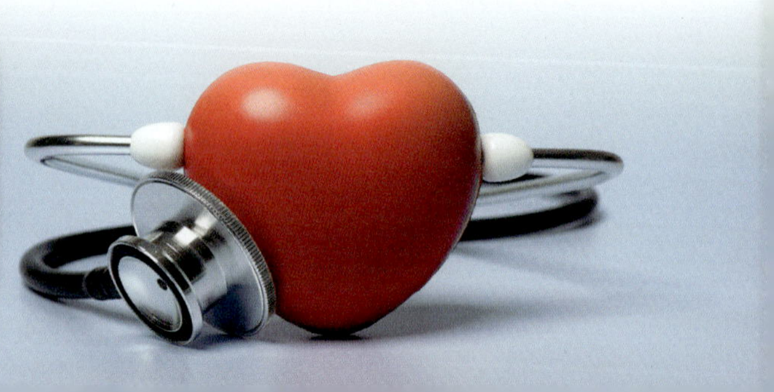

Liquid nitrogen is used for cryopreservation. This is the process of freezing biological specimens for transportation.

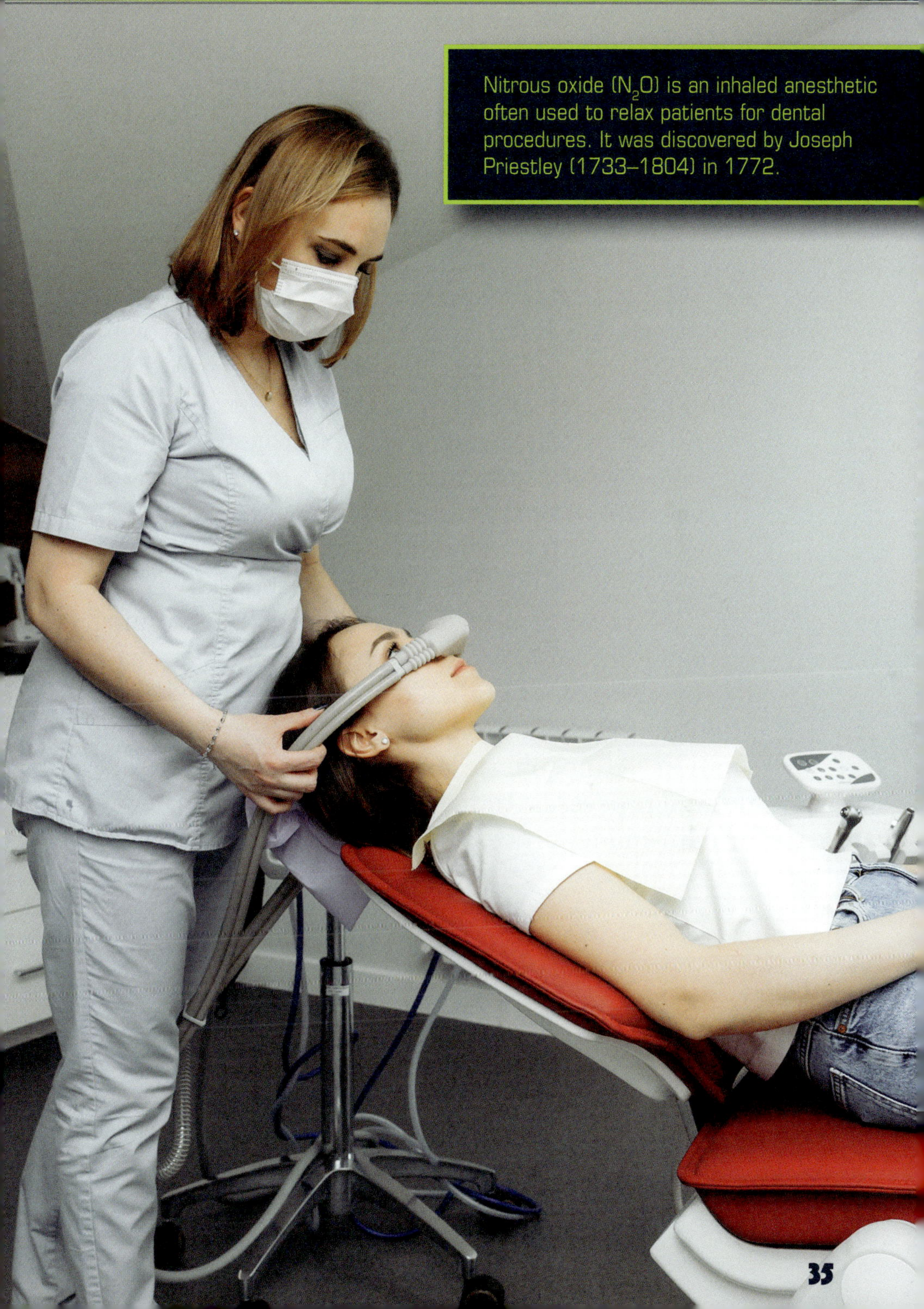

Nitrous oxide (N_2O) is an inhaled anesthetic often used to relax patients for dental procedures. It was discovered by Joseph Priestley (1733–1804) in 1772.

CHAPTER 5

LET'S MIX IT UP!

All living things need nitrogen. Nitrogen is all around us. However, the structure of the nitrogen molecule means it must be broken apart to form compounds that plants and animals can use. The two nitrogen atoms which form a nitrogen molecule (N_2) are connected by strong triple bonds. It takes a great deal of energy to break those bonds. Most living things don't have that amount of energy. So how do they get the nitrogen they need to live?

Nitrogen fixation is the process of transforming nitrogen into forms that living things can use. Nitrogen is generally fixed in one of three ways: atmospheric fixation, biological fixation, or industrial fixation. Atmospheric and biological fixation are natural processes. Industrial fixation is an artificial, or man-made, process.

Most naturally occurring nitrogen fixation is performed by humble microorganisms in the soil. They break the triple bonds.

WHEN LIGHTNING STRIKES

One way to break a nitrogen molecule's bond is to expose it to heat. Heat is a form of energy. When a molecule is exposed to heat, the atoms get excited and move very rapidly. If enough heat is added to a molecule, the atoms will move so much that the bond holding them together will break. The stronger the bond, the more heat, or energy, it takes to break it. Since a nitrogen molecule (N_2) has a very strong triple bond, it takes extremely high temperatures to separate the atoms.

Lightning is nature's way of balancing electrical forces between Earth and its atmosphere. The enormous amount of heat released by lightning is strong enough to break the bonds of nitrogen atoms. That same heat causes the thunder clap you hear seconds after you see lightning.

Lightning provides the kind of heat that is needed to break nitrogen bonds. When lightning flashes, it can heat the air as high as 50,000°F (27,760°C)! After a nitrogen molecule splits through this atmospheric fixation, the nitrogen atoms are free to react with oxygen. When nitrogen and oxygen combine, they form a compound called nitrogen monoxide (NO). This is a colorless gas that is made up of one nitrogen atom and one oxygen atom. Nitrogen monoxide then reacts with other oxygen atoms in the air to form a compound called nitrogen dioxide (NO_2). This compound is made up of one nitrogen atom attached to two oxygen atoms. Nitrogen dioxide is a poisonous, reddish-brown gas with a very pungent odor.

Reactions in the air and soil convert these nitrogen oxides into soluble nitrogen compounds that can be easily absorbed by the roots of plants. Once nitrogen is inside a plant, the plant uses it to make more complex nitrogen compounds, such as chlorophyll. Plants need chlorophyll to undergo photosynthesis. That is how plants make their food. Without fixation, the nitrogen that plants need would not be available.

Nitrogen-fixing bacteria live in the roots of plants.

ALL ABOUT BACTERIA

Biological fixation is the process by which most nitrogen compounds are made. In this process, bacteria in the soil convert nitrogen gas into ammonia (NH_3). Ammonia is a compound made up of one nitrogen atom and three hydrogen atoms.

Once lightning breaks the bonds between nitrogen atoms, they can join with other elements to form compounds that plants need to thrive. These compounds fall to Earth mixed with rain. Animals then eat plants to get the nitrogen they need.

Animals return nitrogen to the soil in their waste products. Bacteria in the soil break down the waste into a form that enriches the soil, helping more plants to grow.

Nitrogen-fixing bacteria are the only living organisms that can use nitrogen directly from the air. Instead of using heat to break the nitrogen bonds, however, these bacteria use an enzyme. An enzyme is a molecule that speeds up a chemical reaction. You can recognize an enzyme because its name usually ends in "-ase." The enzyme that bacteria use to convert nitrogen into ammonia is called nitrogenase. Once the nitrogen bonds are broken by this enzyme, the enzyme helps the nitrogen bond with three atoms of hydrogen, forming ammonia (NH_3). Ammonia in the soil can then easily be converted by other bacteria into compounds that plants are able to absorb. This way, certain plants can get the nitrogen they need.

HELPING MOTHER NATURE

The useful nitrogen compound ammonia can also be made synthetically in a process called the Haber-Bosch process. This process, like atmospheric fixation, uses very high temperatures to break the triple bonds. However, instead of using lightning as a heat source, the Haber-Bosch process uses natural gas. Atmospheric nitrogen gas is combined with hydrogen gas (H_2) at temperatures of about 932°F (500°C) and pressures of about 300 atmospheres. Metal catalysts, which work in a similar way to the enzymes used in biological fixation, help to speed up the reaction.

Chemical fertilizers are used to enrich the soil for farming. The process for synthesizing ammonia was created by two Nobel Prize–winning scientists—Fritz Haber (1868–1934) and Carl Bosch (1874--1940).

Despite the large amount of nitrogen gas in the air, it is often a lack of usable nitrogen in the soil that keeps plants from flourishing. This problem of not having enough nitrogen in the soil can be solved by the ammonia that is produced in the Haber-Bosch process. Ammonia from the Haber-Bosch process can easily be liquefied or combined with other elements and used as a fertilizer for plants. Ammonia is a vital ingredient in fertilizers, and you will see just how important it is when you look at the nitrogen cycle.

THE CIRCLE OF LIFE

The nitrogen cycle describes how nitrogen and nitrogen-containing compounds change as they move throughout nature. From the air, nitrogen passes to the soil, to all living things, and then eventually back into the air. We have already discussed how plants get their nitrogen from the soil through nitrogen fixation, but what about animals? Unlike plants, animals cannot make their own food. Animals must get their nitrogen by eating plants, which have absorbed nitrogen from the soil. Animals also get nitrogen by eating other animals that have fed on plants. The nitrogen locked in the tissues of these plants and animals is released during digestion. Animals then can use nitrogen to form complex nitrogen-containing compounds, such as protein, in their bodies.

The cycle continues when plants and animals die. As dead plants and animals decompose, or break apart, they deposit complex nitrogen compounds into the ground. Certain bacteria that live in the soil break down these nitrogen compounds, turning them back into ammonia. Some of this ammonia is converted by bacteria into soluble nitrogen compounds, which can be reabsorbed by plants. Other bacteria work to turn this ammonia back into nitrogen gas, which is returned to the atmosphere. Because plants and animals are dying and decomposing all the time, the nitrogen cycle stays in balance. Nearly the same amount of nitrogen gas is returned to the air by these bacteria as was taken out by other bacteria. This completes the cycle and keeps the nitrogen content of Earth in a perfect balance.

Nature maintains a perfect balance of nitrogen. Rain falls, plants grow, and animals produce waste and then die and decompose. The amount of nitrogen in the environment remains more or less stable. However, human activities often produce too much nitrogen, upsetting the balance and causing pollution.

WHEN NITROGEN ISN'T SO NICE

We've already discussed the dangers posed by explosives composed of nitrogen compounds such as nitroglycerin or potassium nitrate (a major part of gunpowder). However, a less obvious source of danger is nitrogen pollution. Although the Haber-Bosch process for synthesizing ammonia for use as fertilizer was a tremendous breakthrough for the farming industry, it's possible to have too much of a good thing. While synthetic fertilizers have allowed food production to skyrocket, there's a limit to how much nitrogen crops can use. The remainder ends up in the air and water. Too much nitrogen in water allows algae to bloom excessively. Extra algae use up oxygen that other aquatic life-forms need. They also block sunlight. This can allow bacteria to grow, contaminating the water. People can get sick from drinking the water or eating animals that live in it.

Excess nitrogen can also form harmful gases. Nitrous oxide is a greenhouse gas that's 300 times more powerful than carbon dioxide or methane. Greenhouse gases trap heat, gradually changing Earth's climate. This leads to shifts in weather patterns, including sudden violent storms. In addition, nitrous oxide can remain in the atmosphere for two centuries. Excess ammonia gas and car exhaust (the waste products of combustion) are also harmful. They can damage the atmosphere as well as make breathing difficult for people and animals.

PROTEINS, AMINO ACIDS, AND DNA

Animals—including human beings—inhale nitrogen when they breathe. Airborne nitrogen isn't in a form that can be used, so it's then exhaled. People and animals also absorb small amounts of nitrogen through their skin. However, most of the nitrogen they need comes from eating plants that contain usable nitrogen compounds or eating other animals that have eaten plants. So now that you've ingested nitrogen, what does your body do with it?

The food you eat contains a substance called protein. Protein is a macronutrient that contains nitrogen. Protein is what gives your body structure. Skin, hair, and muscles all contain protein. The organs that keep your body working also contain protein.

When your body digests food, it breaks down protein into smaller structures called amino acids. There are 20 different amino acids that come together in different sequences to form different kinds of protein. Animal protein contains all the amino acids people need. Plants must be eaten in combinations to get all 20 amino acids. The process of digestion breaks down the food you eat into amino acids, then combines them into different sequences, forming substances your body needs to function.

One of these substances is DNA (deoxyribonucleic acid). DNA is a molecule that contains the "blueprint" for your body. DNA molecules are made of even smaller molecules called nucleotides. There are four kinds of nucleotides, each named for the nitrogenous base it's built on—adenine (A), thymine (T), guanine (G), and cytosine (C).

DNA contains the code which specifies the order in which amino acids need to be linked to form different organisms. No two living things have the same DNA. Even identical twins, which might start out the same, experience some genetic changes as they develop.

Nitrogen makes it possible for you to exist. It's in the food you eat, the protein that makes up your body, and even the microscopic structures that make you who you are.

Every single person on Earth has their own, individual DNA code. No two are the same.

Two nucleotides come together to form a base pair. Many base pairs together connect to form the double helix of a DNA molecule. The combinations of nucleotides and the order in which they're arranged form the DNA code—the instructions for a life-form.

CHAPTER 6

LIVING WITH NITROGEN

Often, people think nitrogen is one of those things they learn about in school that doesn't affect them in real life. After all, you can't see or smell the nitrogen in the air, and you can't see DNA unless you have access to a very powerful microscope.

But you *can* see nitrogen in the world around you. Sometimes that's a good thing—nitrogen can be used by chefs and entertainment specialists to produce amazing special effects. Nitrogen gas and nitrogen compounds can be used to construct lifesaving equipment or to help solve crimes.

Unfortunately, sometimes you can see or smell the effects of nitrogen as pollution in the air or water.

POISONOUS POLLUTION

Some of the worst air pollutants are nitrogen compounds. Whenever people burn fossil fuels (such as oil, gas, or coal), nitrogen oxides are produced. Though power plants, schools, homes, and offices all produce nitrogen oxides, automobile exhaust accounts for more than half of the nitrogen oxides in polluted air. The heat from your car's hot engine forms nitrogen monoxide, just like lightning does. When this poisonous gas reacts with oxygen, it forms nitrogen dioxide. This foul-smelling brown gas is one of the main components of smog, the brown haze that is seen hanging over some urban areas. Nitrogen dioxide also dissolves in rainwater and forms nitric acid, one of the ingredients in acid rain. Acid rain kills fish and vegetation and eats away at buildings and statues. Nitrogen dioxide is a common outdoor pollutant, but it can also pollute the inside of your house! It comes from fireplaces, gas stoves, and water heaters. Nitrogen dioxide can irritate your eyes, nose, and throat. It can also make it easier for you to get lung infections. You can keep this gas from building up in your house by opening windows and vents.

Parts of California were known for terrible smog that made people sick. Now the state has the strictest emissions regulations in the country.

WHAT'S IN YOUR AIRBAG?

Car airbags have saved many lives since they were first invented in 1968, but how do they work? An airbag's job is to provide a cushion between the passengers and the car. It does this with the help of nitrogen gas. All airbags are accompanied by a sensor. During a collision, the sensor sets off an explosive reaction that instantly produces hot blasts of nitrogen gas. As you know, gases expand. It is the rapid expansion of the nitrogen gas within the airbag that causes the bag to inflate in a split second, thus saving lives.

Front-seat airbags have been standard safety equipment in cars since the 1990s. According to the National Highway Traffic Safety Administration (NHTSA), they have saved about 50,000 lives. Thanks, nitrogen!

Silver nitrate ($AgNO_3$) is used to collect fingerprints for use in solving crimes. It's also an ingredient in photographic emulsions. Those are the chemicals that capture a photographic image.

SOLVING CRIMES

Whenever you touch something, you leave behind evidence with fingerprints. We are all born with unique ridges on our fingertips. Look closely at your fingers. The patterns you see are unlike those on anyone else's fingertips in the world. Since your fingertips are coated with a film of oily perspiration, your fingerprint is left behind on the surface of the things you touch.

Investigators can use fingerprints to identify people who have amnesia (memory loss), the victims of accidents, and, of course, suspects of crimes. However, fingerprints are often invisible. That is why compounds like silver nitrate ($AgNO_3$) are important. Silver nitrate can be sprayed on surfaces such as paper. Exposure to light causes a reaction that turns the print a brownish color. The print can then be photographed and used for identification.

Cryogenic special effects that use nitrogen are much different from the fog machines available at party stores. They require professional training and handling due to the extremely low temperatures involved.

NOT JUST FUN AND GAMES

Perhaps you've been to a theme park and seen a dragon breathing clouds of smoke or gone to a Halloween event with creepy fog curling around a pretend graveyard. These cool effects can be created with liquid nitrogen. Unlike dry ice (which releases carbon dioxide) or fog machines that run on glycol-based "fog juice" (which can release harmful chemicals such as formaldehyde), liquid nitrogen produces water vapor. Since liquid nitrogen is super cold—about -321°F (-196°C)—these special effects can only be safely created by trained experts. But they are magical to see!

We interact with nitrogen every day. The results are pretty spectacular even if they're not quite so visible. Nitrogen is in the food we eat. Our bodies turn that food into the proteins and amino acids we need to live. Then we return nitrogen to the earth as waste products that enrich the soil. Nitrogen powers the DNA that holds the code for every living thing on Earth.

Of course, human activities can also produce too much nitrogen. Nitrogen pollution contaminates our water and the things that live in it. It also contributes to greenhouse gases and climate change. If we want to preserve the earth for future generations, we need to carefully consider how our actions impact the planet.

Humans can be the earth's worst enemies—or its greatest defenders.

THE PERIODIC TABLE OF ELEMENTS

Nitrogen was included in Mendeleev's original table. Its entry on the modern periodic table includes a lot of information about the element, including its atomic number, atomic mass, and name.

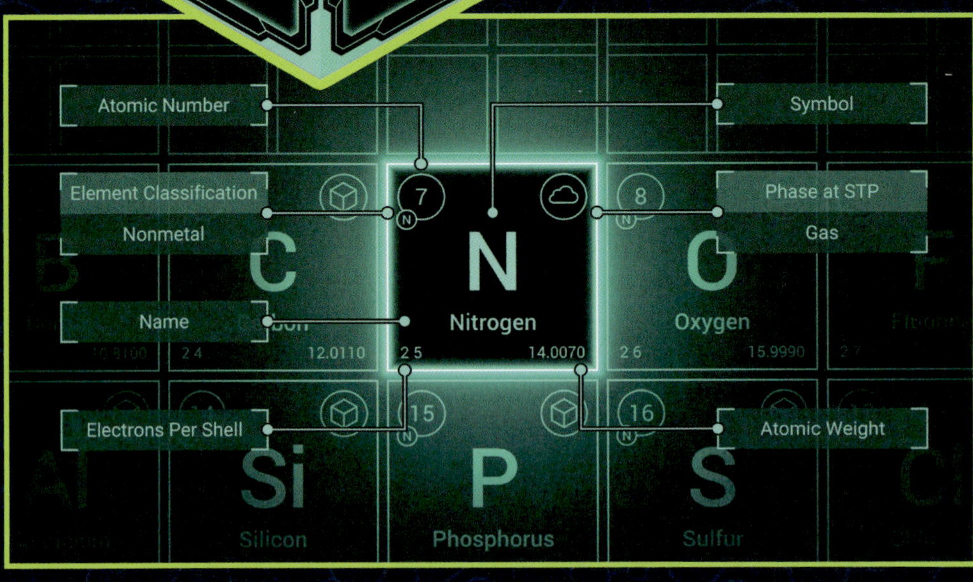

PERIODIC TABLE OF ELEMENTS

- Alkali metals
- Alkaline earth metals
- Transition metals
- Metal
- Metalloids
- Nonmetals
- Halogen
- Noble gases
- Lanthanoids
- Actinoids

1	2	3	4	5	6	7	8	9	10	11	12	13	14	15	16	17	18
1 H Hydrogen 1.008																	2 He Helium 4.0026
3 Li Lithium 6.94	4 Be Beryllium 9.0122											5 B Boron 10.81	6 C Carbon 12.011	7 N Nitrogen 14.007	8 O Oxygen 15.999	9 F Fluorine 18.998	10 Ne Neon 20.180
11 Na Sodium 22.990	12 Mg Magnesium 24.305											13 Al Aluminium 26.982	14 Si Silicon 28.085	15 P Phosphorus 30.974	16 S Sulfur 32.06	17 Cl Chlorine 35.45	18 Ar Argon 39.948
19 K Potassium 39.098	20 Ca Calcium 40.078	21 Sc Scandium 44.956	22 Ti Titanium 47.867	23 V Vanadium 50.942	24 Cr Chromium 51.996	25 Mn Manganese 54.938	26 Fe Iron 55.845	27 Co Cobalt 58.933	28 Ni Nickel 58.693	29 Cu Copper 63.546	30 Zn Zinc 65.38	31 Ga Gallium 69.723	32 Ge Germanium 72.630	33 As Arsenic 74.922	34 Se Selenium 78.971	35 Br Bromine 79.904	36 Kr Krypton 83.798
37 Rb Rubidium 85.468	38 Sr Strontium 87.62	39 Y Yttrium 88.906	40 Zr Zirconium 91.224	41 Nb Niobium 92.906	42 Mo Molybdenum 95.95	43 Tc Technetium (98)	44 Ru Ruthenium 101.07	45 Rh Rhodium 102.91	46 Pd Palladium 106.42	47 Ag Silver 107.87	48 Cd Cadmium 112.41	49 In Indium 114.82	50 Sn Tin 118.71	51 Sb Antimony 121.76	52 Te Tellurium 127.60	53 I Iodine 126.90	54 Xe Xenon 131.29
55 Cs Caesium 132.91	56 Ba Barium 137.33	57-71	72 Hf Hafnium 178.49	73 Ta Tantalum 180.95	74 W Tungsten 183.84	75 Re Rhenium 186.21	76 Os Osmium 190.23	77 Ir Iridium 192.22	78 Pt Platinum 195.08	79 Au Gold 196.97	80 Hg Mercury 200.59	81 Tl Thallium 204.38	82 Pb Lead 207.2	83 Bi Bismuth 208.98	84 Po Polonium (209)	85 At Astatine (210)	86 Rn Radon (222)
87 Fr Francium (223)	88 Ra Radium (226)	89-103	104 Rf Rutherfordium (267)	105 Db Dubnium (268)	106 Sg Seaborgium (259)	107 Bh Bohrium (270)	108 Hs Hassium (277)	109 Mt Meitnerium (278)	110 Ds Darmstadtium (281)	111 Rg Roentgenium (282)	112 Cn Copernicium (285)	113 Nh Nihonium (286)	114 Fl Flerovium (289)	115 Mc Moscovium (290)	116 Lv Livermorium (293)	117 Ts Tennessine (294)	118 Og Oganesson (294)

57 La Lanthanum 138.91	58 Ce Cerium 140.12	59 Pr Praseodymium 140.91	60 Nd Neodymium 144.24	61 Pm Promethium (145)	62 Sm Samarium 150.36	63 Eu Europium 151.96	64 Gd Gadolinium 157.25	65 Tb Terbium 158.93	66 Dy Dysprosium 162.50	67 Ho Holmium 164.93	68 Er Erbium 167.26	69 Tm Thulium 168.93	70 Yb Ytterbium 173.05	71 Lu Lutetium 174.97
89 Ac Actinium (227)	90 Th Thorium 232.04	91 Pa Protactinium 231.04	92 U Uranium 238.03	93 Np Neptunium (237)	94 Pu Plutonium (244)	95 Am Americium (243)	96 Cm Curium (247)	97 Bk Berkelium (247)	98 Cf Californium (251)	99 Es Einsteinium (252)	100 Fm Fermium (257)	101 Md Mendelevium (258)	102 No Nobelium (259)	103 Lr Lawrencium (266)

- C Solid
- Hg Liquid
- H Gas
- Rf Unknown

GLOSSARY

anesthetic: Something of, relating to, or capable of producing anesthesia, or loss of feeling.
atmosphere: The mass of air that surrounds Earth or another planet.
atom: The smallest part of an element having the chemical properties of that element.
bacteria: Single-celled organisms, some of which break down waste and bodies of dead organisms, the products of which will then be reused by other animals.
bond: An attractive force that links two atoms together.
calorie: The amount of heat required at a pressure of one atmosphere to raise the temperature of one gram of water one degree Celsius.
chemical reaction: A change in which one kind of matter is turned into another kind of matter.
covalent: A chemical bond formed between atoms by the sharing of electrons.
diatomic: Consisting of two atoms; having two atoms in the molecule.
energy: The ability to do work (or produce heat).
genetic: Dealing with the variation and makeup of organisms.
homonuclear: Of or relating to a molecule composed of identical nuclei.
kilo-: A prefix meaning 1,000.
mass: The amount of matter in something.

matter: What things are made of, something that takes up space and has mass.
mole: A unit of measurement that contains an exact amount of indivisible units (such as atoms or molecules) of a substance.
molecule: A group of atoms that are chemically bonded together.
phase: A state of matter such as solid, liquid, or gas (and sometimes plasma).
photosynthesis: The process in which green plants make their own food from sunlight, water, and a gas called carbon dioxide.
soluble: Able to be dissolved in liquid.
symbol: Something, such as a letter, abbreviation, or image, that stands for something else.
urban: Of or relating to the city.
valence shell: The outermost shell of an atom containing the valence electrons.
volume: The amount of space that something occupies.

FOR MORE INFORMATION

AMERICAN MUSEUM OF NATURAL HISTORY

200 Central Park West
New York, NY 10024-5102
(212) 769-5000
Website: www.amnh.org
Facebook: @naturalhistory
Instagram and X: @amnh
AMNH is a world-class museum dedicated to advancing research and sharing scientific knowledge about the natural world.

CANADA SCIENCE AND TECHNOLOGY MUSEUM

1867 St. Laurent Boulevard
Ottawa, Ontario K1G 5A3
Canada
(613) 991-3044
Website: ingeniumcanada.org/scitech
Facebook, Instagram, and X: @SciTechMuseum
The Canada Science and Technology Museum is an interactive celebration of Canada's scientific achievements.

CHEMICAL INSTITUTE OF CANADA (CIC)

90-2420 Bank Street
Ottawa, Ontario, Canada
K1V 8S1
(613) 232-6252
Website: www.cheminst.ca/
X: @CIC_ChemInst
The CIC facilitates the research and cooperation of chemists and chemical engineers in Canada.

INTERNATIONAL UNION OF PURE AND APPLIED CHEMISTRY (IUPAC)

IUPAC Secretariat
79 T.W. Alexander Drive
Research Commons Building 4501, Suite 190
Research Triangle Park, NC 27709, USA
(919) 485 8700
Website: iupac.org/
Facebook: @iupac.org
X: @iupac
IUPAC is the global chemistry organization that maintains the periodic table of elements.

INTERNATIONAL UNION OF PURE AND APPLIED PHYSICS (IUPAP)

Fondazione Internazionale Trieste per il Progresso e la Libertà delle Scienze
c/o ICTP
Strada Costiera 11
34151 Trieste
Website: iupap.org/
Facebook: @iupap
X: @iupap_physics
IUPAP's mission is to assist in the global advancement of physics and facilitate international scientific cooperation.

LOS ALAMOS NATIONAL LABORATORY

PO Box 1663
Los Alamos, NM 87545
(505) 667-5061
Website: www.lanl.gov/
Facebook: @LosAlamosNationalLab
Instagram and X: @losalamosnatlab
Los Alamos National Laboratory is a federally funded research and development center dedicated to solving national security issues with science.

FOR FURTHER READING

Congdon, Lisa. *The Illustrated Encyclopedia of the Elements: The Powers, Uses, and Histories of Every Atom in the Universe*. San Fransico, CA: Chronicle Books, LLC, 2021.

DK. *Super Simple Chemistry: The Ultimate Bitesize Study Guide*. New York, NY: DK Publishing, 2020.

McHenry, Ellen Johnston. *The Chemical Elements Coloring & Activity Book: A Fun and Interactive Guide to the World of Chemistry*. Independently Published, 2021.

Emminizer, Theresa. *The Nitrogen Cycle*. Buffalo, NY: Enslow Publishing, 2024.

Hunt, Santana. *The Nitrogen Cycle*. Buffalo, NY: Gareth Stevens Publishing, 2020.

Silver, Donald, and Patricia Wynne. *My First Book About Chemistry*. Mineola, NY: Dover Publications, 2020.

Thomas, Isabel, and Sara Gillingham. *Exploring the Elements: A Complete Guide to the Periodic Table*. New York, NY: Phaidon Press, Inc., 2021.

Zovinka, Edward P., and Rose A. Clark. *A Kids' Guide to the Periodic Table: Everything You Need to Know About the Elements*. Emeryville, CA: Rockridge Press, 2020.

INDEX

ABOUT THE AUTHOR

KATHLEEN A. KLATTE is the author of many nonfiction books for children and teens. Topics she has written about range from animals and nature to constitutional law to unusual career choices. She lives in New York with one cat and far too many books and Legos.

PHOTO CREDITS

Cover, p. 1 remotevfx.com/Shuttestock.com; Cover, pluie_r/Shutterstock.com; Cover, p. 1-64 Oksancia/Shutterstock.com; p.5 Zerbor/Shutterstock.com; p. 6 Ipajoel/Shutterstock.com; p. 7 Jeppe Gustafsson/Shutterstock.com; p. 9 tomertu/Shutterstock.com; p. 11 Maridav/Shutterstock.com; p. 13 Mahir KART/Shutterstock.com; p. 14 Vink Fan/Shutterstock.com; p. 15 Antoine2K/Shutterstock.com; p. 17 7Gen_Grafik/Shutterstock.com; p. 19 YEVHENIIA BUNHA/Shutterstock.com; p. 21 Radzas2008/Shutterstock.com; p. 23 eastern archive/Shutterstock.com; p. 24 IMG Stock Studio/Shutterstock.com; p. 25 Viktoriia Shylling/Shutterstock.com; p. 26 Kondor83/Shutterstock.com; p. 27 Elisa Manzati/Shutterstock.com; p. 28 tunasalmon/Shutterstock.com; p. 29 Sansanorth/Shutterstock.com; p. 30 Cheng Wei/Shutterstock.com; p. 31 LarysaPol/Shutterstock.com; p. 32 Rustam Bikulov/Shutterstock.com; p. 33 WiP-Studio/Shutterstock.com; p. 34 19 STUDIO/Shutterstock.com; p. 35 Inna photographer/Shutterstock.com; p. 36 thodonal88/Shutterstock.com; p. 37 Ross Ellet/Shutterstock.com; p. 38 malshkoff/Shutterstock.com; p. 39 amfroey/Shutterstock.com; p. 40 Pong Wira/Shutterstock.com; p. 41 tawanroong/Shutterstock.com; p. 43 STARSsoft/Shutterstock.com; p. 46 Roman Samborskyi/Shutterstock.com; p. 47 Billion Photos/Shutterstock.com; p. 49 Doidam 10/Shutterstock.com; p. 50 Caito/Shutterstock.com; p. 51 BeataGFX/Shutterstock.com; p. 52 bondart/Shutterstock.com; p. 53 Khanthachai C/Shutterstock.com; p. 54 ShannonChocolate/Shutterstock.com; p. 55 View_From_My_Lens/Shutterstock.com.